Abel Hernández-Muñoz

Safari to the Island of Trinidad

Abel Hernández-Muñoz

Safari to the Island of Trinidad

Chronicle of the trip made by a Cuban ecologist to the jungle of Trinidad & Tobago

ScienciaScripts

Imprint
Any brand names and product names mentioned in this book are subject to trademark, brand or patent protection and are trademarks or registered trademarks of their respective holders. The use of brand names, product names, common names, trade names, product descriptions etc. even without a particular marking in this work is in no way to be construed to mean that such names may be regarded as unrestricted in respect of trademark and brand protection legislation and could thus be used by anyone.

Cover image: www.ingimage.com

This book is a translation from the original published under ISBN 978-613-8-98063-6.

Publisher:
Sciencia Scripts
is a trademark of
Dodo Books Indian Ocean Ltd. and OmniScriptum S.R.L publishing group

120 High Road, East Finchley, London, N2 9ED, United Kingdom
Str. Armeneasca 28/1, office 1, Chisinau MD-2012, Republic of Moldova, Europe
Printed at: see last page
ISBN: 978-620-8-24923-6

TRINIDAD ISLAND SAFARI

A Cuban ecologist studies the island's wildlife for the University of Sancti Spíritus José Martí Pérez in Cuba.

Abel Hernández Muñoz

Scientific Chronicle

INDEX

AN OSSUARY OF ISLANDS

Islands have always been considered magical places, appearing on the horizon when sailors had given up hope of survival. They have been emblematic of the natural exoticism that characterises figurations of earthly paradises. They are places isolated by the sea, subject to it, in total harmony with nature. The stars are best seen from the high continental mountains; however, when one closes one's eyes, one sees them from the shore of a beach on an islet. The sea, which is the cradle of life, has given the islands their best clothes and they are like outcrops of the spirit of the oceans. It is enough to reach them for man to feel the energy they transmit in his whole body. And among the islands, those of the tropics are like the first class of the ship of life, because they are also light and colour. Living on an island, even in the most difficult circumstances, such as those faced by the famous Robinson Crusoe in Daniel de Foe's work of the same name, is a pleasure for the senses. Since the dawn of human history, islands have been the stepping stones of the paths opened in the sea. Without them, Ulysses would not have returned to Ithaca, nor would Leif Erickson have discovered new lands in what is now America, nor would Columbus have been able to reach the New World. In contemporary times, the islands insist on offering their own alternative.

Today, those of us who live on the shores of the Caribbean Sea are struggling to find a link that allows us to reunite with the people of the past, to build a common home that is ours in the landscape that sustains us, alongside the blue that continues to surround us. Our legacy is a mixture of triumph and pain, for the blood of the coloniser, the Indian and the African circulate together in most of us. For some, the Spanish conquistadors were once their masters, for others the French, in other cases the British or the Danes. Today we struggle to find a unity that embraces us all.

In the 1950s, when several Caribbean territories were seeking political independence from British colonial rule, the debate over the creation of a Federation of the West Indies was much more enthusiastic. A flag was designed, a capital was named and the Caribbean leaders of the day emulated each other to determine who would be the region's new prime minister. An anthem with words and music was even composed.

The federation was short-lived and the regional anthem became that of Trinidad and Tobago, but hidden in its music was a phrase that still recalls the longing for the unity of our common heritage:

"...We stand shoulder to shoulder,

Islands of the Blue Caribbean Sea..."

So when I found out through an office with Jorge Luis Aneiros

Alonso, rector of the University of Sancti Spíritus José Martí Pérez, that the Association of Caribbean States was organising a literary competition for Caribbean writers who wanted to write about the Caribbean and selected me, a creator of popular science, to write a book about this endearing geographical area, I was overcome with a tremendous wave of enthusiasm, because the island territory selected was none other than the Republic of Trinidad and Tobago. This is not just any country in the West Indies, but the closest to the mainland, and guess what, it has jungles, so for the first time I would be visiting the jungle.

Once the meeting was over, I left the Rector's office and went to the Navigation Room, I found myself there, sat down in front of one of the machines, entered my account password and once ready, I clicked on the Google search engine, typing in its search bar five words: Republic of Trinidad and Tobago, instantly several documents on the subject appeared on the screen but I selected the Wikipedia one, it opened and I could read the following:

"**Trinidad and Tobago**, also called **Trinidad and Tobago**, [1]whose official name is the *Republic of* **Trinidad and Tobago**, is one of the thirteen countries that make up the Insular America, West Indies or Caribbean Islands, one of thirty-five on the American continent. Its capital is Port of Spain and its most populous city is Chaguanas.

It is located at the southern tip of the West Indies, south of Grenada, and on the continental shelf off the east coast of Venezuela. With 1 349 667 inhabitants in 2015, it is the sixth most populous country in the Caribbean, behind Cuba, the Dominican Republic, Haiti, Puerto Rico and Jamaica.

Its territory is made up of two main islands, Trinidad - the largest and most populous - and Tobago, plus several smaller islands, and is organised into fourteen corporate regions. Its form of government is a parliamentary republic.

Before the arrival of the Spanish in the 15th century, the island now called Trinidad was populated by Caribs (Nopoya and Supoya), while Tobago was populated by the Kalugo. The indigenous name for Trinidad was Kairi or Leré, which would have meant "Land of Hummingbirds" or perhaps simply "The Island". Christopher Columbus discovered the main island on 31 July 1498 and called it "Tierra de la Santísima Trinidad", while he called the island now called Tobago "Bella Forma". The islands were later disputed by the Spanish, English, Dutch and French, and there was even a colony of Latvians from Curlandia.

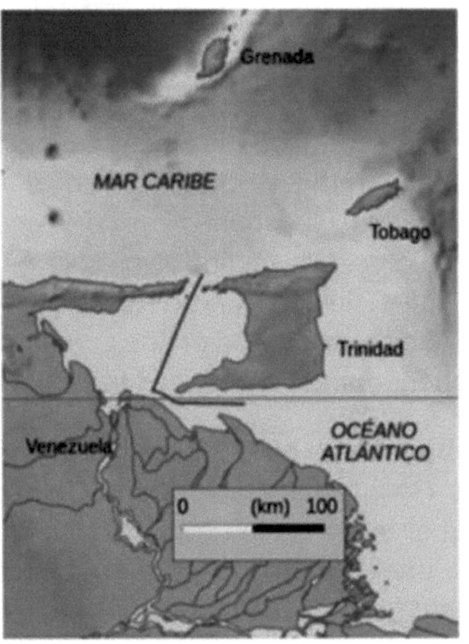

In 1592 Antonio de Berrio established the first permanent enclave, San José de Oruña. *Sir* Walter Raleigh, in search of "El Dorado" in South America, landed in Trinidad on 22 March 1595 and attacked San José de Oruña, capturing both Berrio and the cacique Topiawari.

The Province of Trinidad was created in the 16th century by the Spanish and its capital was San José de Oruña. At the end of the 18th century it was under Spanish control as part of the Captaincy General of Venezuela, but during the Napoleonic Wars, in February 1797, a British force began to occupy the territory. In 1802, through the implementation of the Peace Treaty of Amiens, the islands of Trinidad and Tobago passed to the United Kingdom. France, which also had claims to the territory, formally ceded its claims to the UK in 1814.

The indigenous population virtually disappeared, being largely replaced by the Melano-African population forcibly introduced by the British as slave labour for the sugar cane and tobacco plantations. During the 19th century, as slavery became less economical for Europeans, the immigration of *coolies* from India and China was encouraged, The same was true of Portuguese immigration, while French refugees from Haiti before the Peace of Amiens were better off, although they did not reach the status of the British minority.

In 1899 Tobago, which had been administered from Barbados, became administered from Port of Spain in Trinidad. During World War II, the United States, an ally of the UK, established a major military base on Chaguaramas. Both islands were ruled as a British crown colony for a decade.

On 3 January 1958, Trinidad and Tobago together with Jamaica, Barbados, the Windward and Leeward Islands and other smaller islands formed the Federation of the West Indies, with Chaguaramas as its federal capital. This federation disappeared in 1961 when Jamaica seceded and independence was granted (albeit with the British monarch as head of state) on 3 August 1962. Throughout this period there were serious ethnic conflicts, mainly between people of African descent and people of Indian descent.

Trinidad and Tobago joined the UN that year and joined the OAS in 1967 and the OAS in 1976. Trinidad and Tobago formally declared full independence from the British crown, although it has remained within the Commonwealth of Nations.

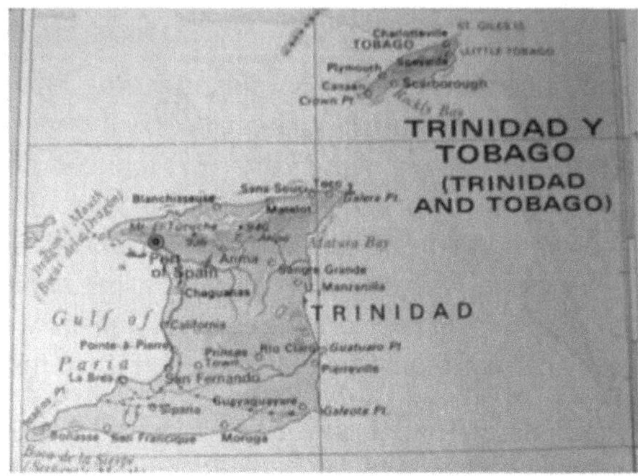

Trinidad and Tobago is an archipelago consisting of several islands in the Southern Caribbean, between the Caribbean Sea and the North Atlantic Ocean, northeast of Venezuela. They are islands of the Lesser Antilles, located on the mainland of Central America and the Caribbean.

Covering an area of 5,128 km^2 , the country consists of two main islands: Trinidad and Tobago, and several smaller islands. Trinidad is 11 km off the northeast coast of Venezuela, and 130 km south of the Grenadines. The island is 4768 km2 (or 93% of the country's total area). The island is quadrangular in shape, with three peninsulas rising from it. Tobago is 30 km northeast of Trinidad and has an area of about 298 km2.

Geologically, the islands are not part of the West Indian archipelago. Indeed, Trinidad was formerly part of the mainland portion of South America and is located on its continental shelf, and Tobago is part of a sunken mountain range connected to the mainland. The islands are now separated from the mainland by two straits, the so-called Dragon's Mouth to the north of the Gulf of Paria and to the south the Columbus Channel.

Trinidad is crossed by three different groups of mountains that are a continuation of the Venezuelan coastal mountain range. The highest, El Cerro del Aripo, the other, El Tucuche. The Central Range runs diagonally across the island and is a low range with marshy areas. The Caroni Plain, composed of alluvial sediments, extends southwards, separating the Northern Range from the Central Range. The southern range consists of a line of hills with a maximum elevation of 305 metres, in a depression left by an ancient volcanic crater there is a curious tar lake.

The coasts of Trinidad are relatively flat and, despite the latitude and the warmth of the sea water, the almost total absence of living coral reefs is striking. This can be explained by the fact that the mud from the Orinoco basin settles on the coasts, and if this mud prevents the formation of corals, it facilitates the existence of an exuberant coastal flora and fauna, among which the four-eyed fish stand out (in fact, they have two eyes divided horizontally in half so that the lower half can see underwater and the upper half can see above water level), tamanduas, howler monkeys and howler monkeys can see above water level: They have two eyes divided horizontally in half so that the lower half sees underwater and the upper half of the eyes sees above water level), tamanduas, howler monkeys and spider monkeys, which are fauna that does not exist in the other Antilles.

There are numerous rivers and streams on Trinidad Island, the most prominent being the 5-kilometre-long Ortoire River, which runs eastward into the Atlantic, and the 40-kilometre-long Caroni River, which flows into the Gulf of Paria. Most of Trinidad's soils are fertile, with the exception of the unstable soil found in the southern part of the island.

Trinidad and Tobago, which lie between the tropics, enjoy a tropical maritime climate influenced by northeasterly winds. In Trinidad, the average temperature is 26°C, and the average maximum temperature is 34°C during the day and an average of 20°C at night. Humidity is high, particularly during the wet season, when it averages 85%.

The island receives an average of 2 110 mm of rainfall per year, generally concentrated in the months of June to December, when, in short, heavy downpours occur frequently. Rainfall increases in the Northern Range, where it can receive up to 3 810 mm. During the dry season, droughts strike the upper central part of the island of

Trinidad".

TRIP TO THE ISLAND OF TRINIDAD

I finally arrived on the island of Trinidad in July 2016. One of my biggest childhood dreams had come true. I had always been excited to think about the jungle and the fascinating flora and fauna that live there, and there it was: the very jungle of the Orinoco River Delta in an island context!

For many people, it would have seemed impossible for me to achieve these goals. Let me tell you how I came to the forest of Trinidad and how chance granted me this great dream.

Since that memorable year I focused my attention in this direction. As I mentioned at the beginning, I had been very interested in the jungle since I was a child. Despite my physical limitations, the desire was greater than ever, especially when I learned that on this small island, smaller than the province of Sancti Spíritus, there were real rainforests on the plains.

As luck would have it, we were invited by the Trinidad and Tobago Writers' Association, who would fund our stay with a view to our writing a book about the Republic of Trinidad and Tobago.

We left for Havana, arriving at the National Bus Terminal at 12:45 pm, and once there we went to the UNEAC headquarters at 17th and H, where we stayed for the night.

The next day we woke up at 4:00 am, and after getting cleaned and dressed, we boarded a taxi to Terminal Three of the International Airport "José Martí", at 8:30 am, to board the Boing-747 of the Constelation Group of the airline Copa Airlines, with destination Panama City.

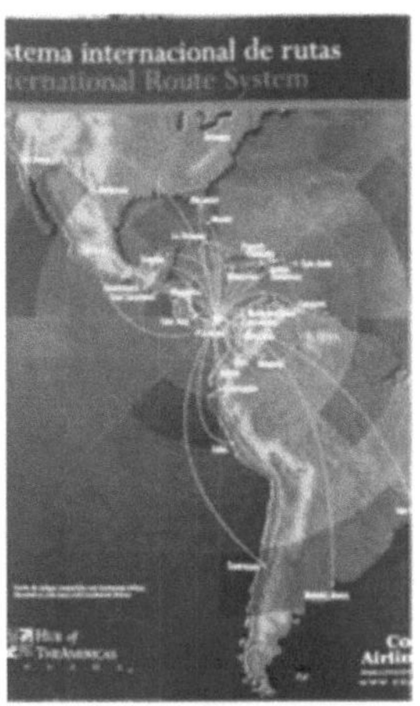

And in the plane, I sat in the window seat, from where I could watch the details of the aircraft's take-off in detail. The big metal bird taxied lazily towards the tip of the runway and, once there, began to gradually pick up speed. Through the video screen located on the back of the front seat and through the central audio system, the crew informed us of the aircraft's characteristics, invited us to fasten our seat belts and told us what we should do in the event of an air disaster.

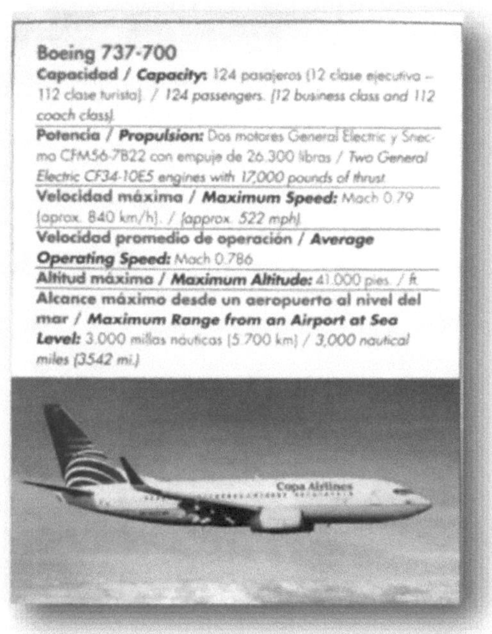

Boeing 737-700

Capacidad / *Capacity:* 124 pasajeros (12 clase ejecutiva – 112 clase turista). / *124 passengers. (12 business class and 112 coach class).*

Potencia / *Propulsion:* Dos motores General Electric y Snecma CFM56-7B22 con empuje de 26.300 libras / *Two General Electric CF34-10E5 engines with 17,000 pounds of thrust.*

Velocidad máxima / *Maximum Speed:* Mach 0.79 (aprox. 840 km/h). / *(approx. 522 mph).*

Velocidad promedio de operación / *Average Operating Speed:* Mach 0.786

Altitud máxima / *Maximum Altitude:* 41.000 pies. / ft.

Alcance máximo desde un aeropuerto al nivel del mar / *Maximum Range from an Airport at Sea Level:* 3.000 millas náuticas (5.700 km) / *3,000 nautical miles (3542 mi.)*

The big metal bird was getting faster and faster, the suburban landscape around the air terminal was passing before my eyes in a dizzying way, until suddenly the plane jumped and was suspended in the air of the capital of all Cubans. Seconds later we could see the agricultural crops, the buildings and the means of transport of Havana, which from that height looked like toys. Then it was the Caribbean blue that added its colour to the already diverse palette of colours. It impressed my senses, the watery infinity of the sea and the gaseousness of the sky, extending towards our southern destination.

Next, the boats looked like little paper boats, and the endless stream of clouds below the wings of our aircraft. The cloud masses seemed to us like figures of animals and objects of different shapes.

The hours passed and we were already flying over the Gulf of Darien, so masterfully described by Emilio Salgari in his work "The Black Corsair". Moments later the imposing mass of the few tall buildings of Panama City, the famous inter-oceanic canal seen from a bird's eye view. Later, the opposite manoeuvre to take-off. The

Jumbo turned gracefully to the right and downwards, looking for the runway and, once on it, deployed the flaps and dropped heavily on the tarmac, with all its weight on the landing gear. It then looked for the chute at the appropriate gate of the air terminal, where we were to descend, and there it stopped its engines.

The airport is spacious and it seems that in every small place there is a business, the facilities support a large flow of passengers arriving or departing to the most dissimilar destinations.

We entered a café in search of sustenance to continue our journey, the pipe tables with their attached chairs painted bright red. Julio Miguel Llanes López, the other guest writer, treated me to pizza and coffee.

- Is there anything else you want? he said.
- Just a glass of water. I replied.
- Do you have an appetite?
- No, it's just that my excitement is such that I'm not even hungry. Later we went to the waiting lounge corresponding to our boarding gate of the Embriager that would take us to Port of Spain. A medium-sized aircraft, not as luxurious as the Boing-737, but very modern and safe. Besides, it is a Brazilian aircraft and that was

already an added value.

There we sat down and had a lively chat, started by Llanes as usual, characteristic of a great conversationalist.

- You're going to get your boots on, boy!
- Of course! This is a unique opportunity to visit the South American rainforest in its Orinoco and insular variant.
- That will be better for you than any postgraduate degree.
- Of course, I know that the experiences there will not be surpassed by any other experience before or since.
- How nice, you deserve it!

So lively was our conversation about our plans that time ran out and we were startled when we heard the announcement in perfect English with a British accent over the audio system of the airport terminal, announcing the departure of our flight to Piarco. We jumped up abruptly and stood last in the queue of passengers, who were engaged in the routine check of travel documents.

Amused as usual, I came back to reality, when the beautiful girl said to me: - Me I help you?, - Yes, I replied offering her my passport and ticket. She looked at my documents and said, - Ockey, see you to five minuts. -Thankyou very much," I replied with a wide smile, and she mixed mine with her chocolate gesture, revealing a perfect set of teeth.

So, I headed for the boarding gate, when Julio was already walking down the sleeve towards the plane. I did the same and once there, my friend Llanes and I had a brief tussle, as I decided to impose myself on the desire to sit on the side of the window.

- Hey, you're going to put the cannon in me again!
- Yes, my brother, you know how important it is for me to be able to contemplate the aerial view of the region.
- All right,...all right... don't protest any more, take the window.

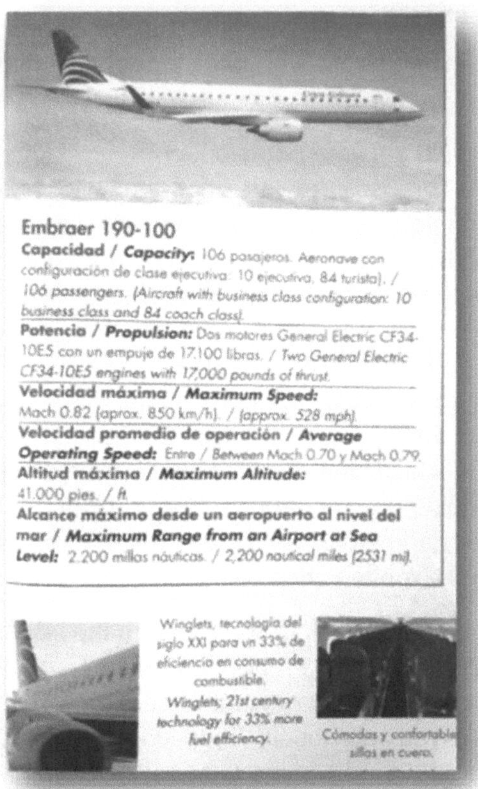

Embraer 190-100

Capacidad / *Capacity:* 106 pasajeros. Aeronave con configuración de clase ejecutiva: 10 ejecutiva, 84 turista). / *106 passengers. (Aircraft with business class configuration: 10 business class and 84 coach class).*

Potencia / *Propulsion:* Dos motores General Electric CF34-10E5 con un empuje de 17.100 libras. / *Two General Electric CF34-10E5 engines with 17,000 pounds of thrust.*

Velocidad máxima / *Maximum Speed:* Mach 0.82 (aprox. 850 km/h). / *(approx. 528 mph).*

Velocidad promedio de operación / *Average Operating Speed:* Entre / *Between* Mach 0.70 y Mach 0.79.

Altitud máxima / *Maximum Altitude:* 41.000 pies. / *ft.*

Alcance máximo desde un aeropuerto al nivel del mar / *Maximum Range from an Airport at Sea Level:* 2.200 millas náuticas. / *2,200 nautical miles (2531 mi).*

Winglets, tecnología del siglo XXI para un 33% de eficiencia en consumo de combustible.

Winglets; 21st century technology for 33% more fuel efficiency.

Cómodas y confortabl sillas en cuero.

As the aircraft took off, I was surprised to find that Julio was fast asleep, while I was delighted to gaze through the glass at the rugged mountains of Colombia and Venezuela.

The rivers were little silver threads and the mountains were chocolate cones, covered with mint. The villages, towns and hamlets resembled sets of little coloured boxes. Those landscapes reminded me of the geography teaching models when I was a seventh grader in Cabaiguán, my childhood village in Cuba.

In this unstable weather, with the sky infested with small local storms, which generated great turbulence for our aircraft, which was flying at eighteen thousand feet (about ten kilometres altitude) and a speed of eight hundred and fifty kilometres per hour, the pilot was forced to make sharp dive descents, which woke Julio up and made him remember his late mother several times.

At last we reached the Gulf of Paria, an enormous emerald arm of the sea, at the north-easternmost point of which, opposite the Orinoco delta, I discovered the islands of Trinidad and Tobago. Later, the island of Trinidad, which, as if in the zoom of a camera, was approaching us in all its greenery and unlikely physiography.

I can already see the international airport, with the grey ribbon of its only runway! We have arrived!

WE ARE ALREADY IN PIARCO

Flying over the airspace of Trinidad, I notice the presence of the Piarco air terminal, with the grey tape of its only runway. A few minutes later we land on this island of enchantment, where the Caribbean is even more charming. Once again, the big runway, the belly of the Embriaer-190, which looks like it is about to hit the pavement, while the ailerons unfold and the landing gear comes out like a surprise doll.

The first dizzying and then slow journey of the aircraft until it is taxied to the corresponding apron of the aforementioned air terminal.

This is Piarco, the great airport of the Insular Caribbean, huge and modern, through its glass walls and ceiling, through its circular main lobby, we discover the soul of these islands and their turquoise blue skies. It is the domain of the most modern technology, but also a showcase of ethnicities and cultures. All civilisations converge and mingle here, an incredible tropical concert of landscapes and sonorities. It is a blessed place, where any Caribbean person would like to be, as a portion of our greater Caribbean homeland. You find yourself in a foreign country and yet it seems to you that there is something of Cuba, Jamaica or Haiti there.

It is considered one of the most modern airports in the insular Caribbean. The airport was opened on January 8, 1931, to serve the

"Compagnie Generale Aeropostale" of Venezuela. Prior to this, the Sabana de Parque de la Reina, the Mucarapo Field, and the Cocorite Port (for seaplanes) were used as airstrips to serve the island. A major extension of the airport, which included the construction of a new terminal building, and high-speed taxiways, was completed in 2001. The old airport building is currently used for cargo handling. The airport is the primary hub and base for Caribbean Airlines.

At Piarco International Airport there are two high-speed taxiways and three connector taxiways (ICAO Code F for new large aircraft). Technologically the airport has 82 state-of-the-art ticketing positions operating under the SITA fibre optic C.U.T.E. system, which exceeds ICAO and IATA recommended standards. It also has a Flight Information Display System, which serves all airport users and a Baggage Information Display System.

The passenger terminal is air-conditioned and smoke-free. It is equipped to handle a peak hour passenger traffic of 1,500 passengers, handled by a fully automated and user-friendly immigration system, which minimises long queues and passenger inconvenience. This convenience is also found in the customs hall, which has loading trolleys for large baggage.
A manager/operator building for the Trinidad and Tobago Air Guard is a builder at Piarco Air Base. Also, a military airfield will be built near the air base.

The new North Terminal consists of 35 964 square metres (380 000 square feet) of building with total layout consisting of three main elements: a main ground structure, a single-level duty-free shopping mall, and a two-level 'Y' formed confluence.

One hundred feet (30 m) of cathedral ceilings and glass in the walls, provide passengers and other visitors at the North Terminal with a sense of open space and magnificent (panoramic) views of the Piarco savannah and the nearby Sierra del Norte mountain range.

The public atrium has the largest glass dome on Caribbean premises. The airport is large enough to accommodate large aircraft such as the Boeing 747, Boeing 777 and Airbus A-340.

Piarco International Airport has a layout consisting of the terminal, the main building which includes three concourses. These concourses are not strictly identified as their name represents, but are divided into the following areas; Gates 1-7, Gates 8-14 and the Tobago, the concourse serving flights into Tobago.

The Trinidad and Tobago Air Guard is based at Piarco International Airport, a modern military force at the service of the Trinidadian people.

A new control tower is being planned. It will have the most modern radars as opposed to the existing old tower.

The voice of Julio Miguel LLanes López pulls you out of your entertaining reading:

- My boy, here we part company. I am still in the capital, as I have to give a series of lectures at the West Indies University, while you are to leave with Mr. Henry for the William Beebe Tropical Research Station.
- Llanes, I wish you every success with your conferences.
- And may you have the most fascinating adventures on your trip to the jungle!
- Thank you, you know that's what I want most.
- Well son, see you in a week, for our return to Cuba. Bye!

Abel boarded the bam, bound for the William Beebe Tropical Research Station.

TRINIDAD, THE MAGIC OF AN ISLAND

I recently had the good fortune to visit Trinidad, the island that the celebrated naturalist Beebe used as his odd-field polygon, Port of Spain is the capital, but the spirit of a city, even though it is the largest in the country, is blurred. It's a bewildering mix of progress and the feeling of being for moments in the middle of a village, where modernity surprises.

There are not many buildings, although there are some, and they are colourful, but most of them are mainly dedicated to commercial or government activities such as: the Eric Williams Financial Complex, the Convention Centre, the government headquarters, the National Library, the National Museum and Art Gallery.

People live in houses, which descend from "the heights" towards the sea, whose architecture is at times defined and at the same time exudes an air of contemporaneity. The day surprises you with thousands of zinc gable roofs that resemble a postcard, albeit a sui generis one.

Because Port of Spain seems to have grown from within and you are surprised to see how, alongside the houses, the industrial zone grows: a thermoelectric plant, a port, a factory, any business associated with the oil and natural gas industry - the mainstay of its economy and that of the whole island - can appear as you walk through the heart of the city.

The Republic of Trinidad and Tobago is an archipelago of two main islands, the largest of which is Trinidad, off the coast of Venezuela. Trinidad was discovered in 1498 by Christopher Columbus, then colonised by the Spanish, but in the 17th century it began to be attacked by the French, English and Dutch who wanted to take over this important enclave. It was the English who succeeded, but it was not until 1802, thanks to the Treaty of Amiens, that ownership of the archipelago was recognised.

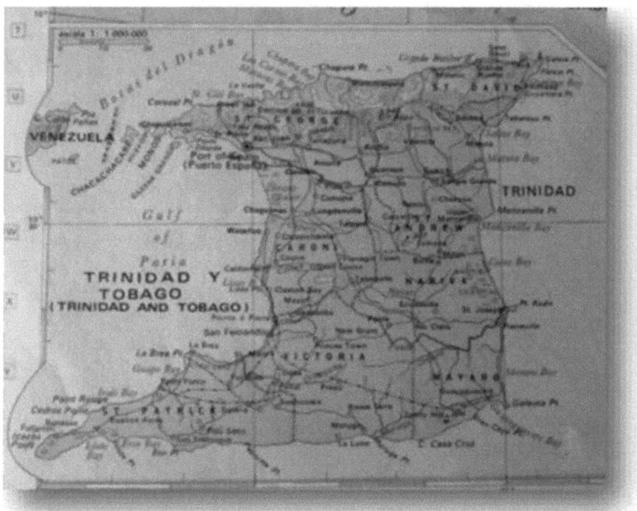

Administratively it is divided into eight counties and has been an independent republic since 1 August 1976. The head of state is the President of the Republic.
The largest religious group in the country is Catholic with 370,000 followers, followed by Protestant with 283,562 followers and at a considerable distance other smaller groups such as Hindus and Muslims.

Education is free at the primary and secondary levels. Higher education is provided by the University of St. Augustine (since 1960) and some colleges. The official language is English.

Between the Gulf of Paria, the hills of the Northern Cordillera and the Caroni swamp - on the island of Trinidad, which is larger and more populated than Tobago and the rest of the small islands that make up the archipelago nation - is born this city that is said to have won part of the land from the sea, and is home to the largest banks in the Caribbean, and one of the most important maritime transport centres in this area.

Port of Spain, I believe, barely retains from its years as a Spanish colony the names of the regions, streets and houses it inherited, such as the Chacón street, which is said to be the memory of the last Hispanic governor. Because this country was deeply marked by the almost two centuries of British colonisation.

Here Trinidadians spend the year preparing for an event that brings everything to a standstill every twelve months, the carnival, which is one of the oldest, largest and most lavish in the world and one of the most important festivities in the country in the annual cycle of its culture. It is the time when the greatest number of tourists arrive on the island and where the Soca, an Afro-Caribbean rhythm born of Calypso, reigns supreme. The parade ends in the Gran Sabana de la Reina, which is in the centre of the city and is a field bigger than two football stadiums. It is said that its wealthy owner donated it to the town on the condition that nothing would be built on it. And so it is.

There is no comprehensive public transport system. There is no railway and a bus is rare. The closest thing to it are private minibuses, known locally as mini-taxis, which cover the east, west and south corridors of the city.

If you want one of the best views, just get to Fort George, some 1,700 feet above sea level, from where on clear days you can make out the mountains of Venezuela to the west. First called La Vigia, it was part of a complex of fortifications to protect against the Napoleonic fleet, but it was never completed and its construction was abandoned as the largest defensive position on the island of Trinidad, although it never saw any military action.

Trinidadians say that you have to come to Port of Spain and Trinidad & Tobago to talk about the Caribbean. They also say that there is no shortage of components to the mix that inhabits these islands, with Afro-descendants, Hindus, Muslims, Canadians, Portuguese, English and Latin Americans.

There are "Twin Towers" in Trinidad. That's what Trinidadians call the Eric Williams Financial Complex, located in Independence Square in Port of Spain.

And there is no taller building in Trinidad and Tobago, or in the entire English-speaking Caribbean, than this pair of 22-storey, 302ft (92m) high skyscrapers. The first tower of the complex houses the country's Central Bank, while the second tower houses the Ministry of Finance.

Eric Williams Place is officially named after the nation's former Prime

Minister (Eric Eustace Williams) and also a noted historian and founder of the People's National Movement (1955).

There is another building with the same name, the Eric Williams Medical Sciences Complex, which functions as the University Hospital of the Republic, associated with the University of the West Indies (West Indians University).

Finally, I do not want to overlook our visit to the National Museum, located in a centenary building, it consists of two exhibition floors, the first dedicated to the history of the main industries of the country, highlighting among them the oil and gas extraction, while the first floor has a panoramic exhibition that starts from the nature of the territory to then address the aboriginal stage, followed by the colonial, and finally the ethnic groups that make up this colourful nation, all displayed through a series of dioramas, showcases, gigantographies and auxiliary texts of exquisite workmanship.

AN ENCOUNTER WITH THE LEGACY OF WILLIAM BEEBE

But if a visit to the Museum is exhilarating, a trip to Simla is even more extraordinary. What a thrill it is to visit the field station of the great naturalist whose name now bears the William Beebe Tropical Research Station (WBTRS), located in the Arima Valley of the Northern Range of Trinidad, founded by Dr. Beebe in 1949 as a Tropical Research Station of the Zoological Society of New York. The Doctor renamed it "Simla" after his campaign through the Indian portion of the Himalayan Mountains during which he published the monumental work "A Monograph of the Pheasants". In 1950 he donated the site to the New York Zoological Society, now the Wildlife Conservation Society.

Beebe was an American scientist, adventurer and writer, and is considered "the father of Neotropical Ecology". He first visited Trinidad in 1908 and fell in love with the wildlife of this land.

The Zoological Society of New York sponsored the station until the 1970s, during which time the station developed important studies of neotropical ecology, botany, entomology, ornithology, herpetology, as well as bats, fish and crustaceans. Many of these papers were published in *Zoologica,* the Society's scientific journal. A total of 306 articles were published based on the work carried out at the field station during the more than 50 years of its existence.

Dr. Beebe died at Simla in June 1962 and Dr. Marc Buchanan continued as Director until 1970, when the field station was closed due to inability to meet maintenance costs. A period of dormancy followed until Simla was donated to the Asa Wright Nature Center (AWNC) in 1974, at which time the AWNC has kept the field station active to facilitate tropical research.

The WBTRS is operated by the Asa Wright Nature Centre, a trust set up for environmental education and conservation. All revenues are used to cover the operating costs of the Station. The AWNC subsidises the scientific research and Simla provides the logistics at low cost and ensures the support of a research assistant at all times. In addition AWNC assumes the annual capital deficit required by Simla.

The William Beebe Tropical Research Station is located in the Arima Valley of the North Range of Trinidad Island, at a latitude of 10^0 41' 1" N and longitude of 61^0 17' W. Trinidad Island is situated in the Orinoco River delta and biologically speaking is South American, it was connected to the mainland 11 000 years ago before the present. The Arima Valley is located approximately in the central portion of Trinidad's North Range and is irrigated by the Arima River as well as the Caroni River to the south.

Located at about 800 feet (approximately 240 m) elevation, the WBTRS receives 2 590 to 3 070 mm of rainfall annually, and its climate is considered tropical. Seasonally there is a dry period from January to May and the rainy season is from June to December. The field station has a land area of 211 acres with good national access. In addition, more than 1,000 acres attached to the AWNC are also available for research activities.

The vegetation of the Research Station includes a matrix of seasonal evergreen forest, lowland rainforest (tropical rainforest) and cocoa plantations in various stages of succession (Beebe, 1952). A large proportion of the wildlife species known from the island have been recorded on Simla (e.g. 164 bird species, 27 snake species and 16 amphibian species). In addition, this central location in the Northern Range has comparatively easy access to other key habitat types on the island.

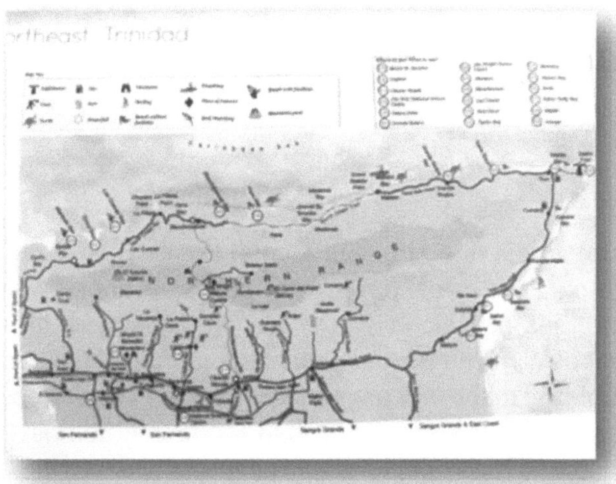

BEFORE GOING TO THE JUNGLE

I knew from the scientific literature that tropical forests grow in a large equatorial belt where temperature, rainfall and day length hardly vary from season to season. The combination of warm temperatures, lots of rain and constant day length contribute to an environment in which plant growth and reproduction is basically independent of the time of year, so that leaves, flowers and fruits are always present to feed the animals.

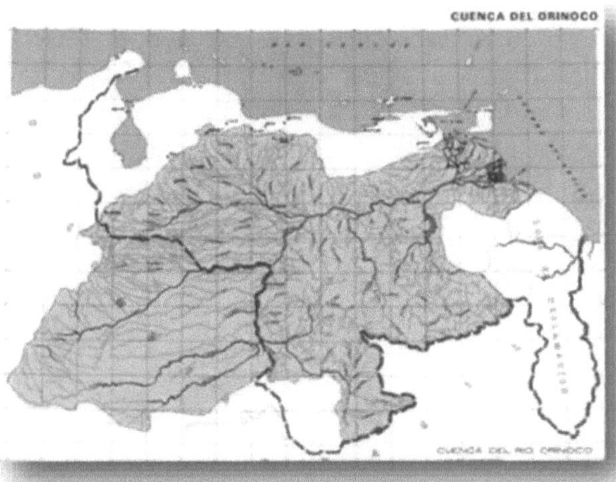

The dense rainforest is a stratified jungle: the crowns of rainforest trees form a continuous canopy (upper stratum), above which the tops of giant trees occasionally protrude. An initial floor of smaller trees and woody climbing plants, or lianas, produce a jungle of such internal complexity that many animals, even some large ones, rarely or never descend to the ground. Within the forest, competition for light is intense and many plants manage to get close to the sun by climbing up the large trees or growing as epiphytes on their branches and trunks. Medium-sized trees form a fairly continuous intermediate stratum.

Despite the myth of the jungle as an impenetrable tangle of vegetation, the soil is so dark that few plants grow in comparison to other types of natural region, except in the sunny clearings created when a tree is blown down by wind and landslides. The microclimate is unfavourable, for although the decomposition of plant matter increases the rate of carbon dioxide, oxygen is in short supply, which is an obstacle to respiration. Because of the rapid recycling of nutrients back into the vegetation, forest soils are rather poor.

Although the diversity of forest species is legendary, many of the trees are superficially quite similar: leaves that shed water from their sharp tips are characteristic of many unrelated species; because there is no need for a thick, impermeable canopy in the forest, the bark tends to be thin; buttressed, buttressed roots with buttresses can enhance stability in shallow or waterlogged soils. As distance from the equator increases, seasonal differences become more pronounced and the number of deciduous species grows, but since each species drops its leaves at different periods and leaves are constantly renewed, the resemblance between semi-evergreen and evergreen forests is close.

The Amazon Basin rainforest is often regarded as the quintessential rainforest and is, in fact, extraordinarily diverse in species of fauna and flora, most of which are as yet unclassified. The Amazon rainforest block is the largest on Earth and, despite population intrusion in many places, remains relatively intact. Also in the Orinoco River basin and especially in its delta, extensive and intricate jungles develop. Alongside these continental jungles, patches of Caribbean rainforest still cover parts of the West Indies.

Rainforest is defined as a forest formation characterised by lush vegetation and relatively high temperatures and rainfall throughout the year. Rainforests are the most biologically diverse ecosystems in the world. Although they occupy less than 7% of the surface of emerged land, they contain more than 50% (some scientists say more than 90%) of the world's plant and animal species. One hectare of tropical rainforest can contain more than 600 tree species. By way of comparison, the forests of the United States and Canada combined have only about 700 tree species. The number of animal species found in rainforests is even greater. In one study, more species of ants were found on a tree trunk in a rainforest than in the entire British Isles.

To classify a forest as rainforest, it is necessary that the treetops of its trees touch and intertwine with each other, forming a dome of vegetation that hardly allows light to pass through. In addition, it must have high temperatures and abundant rainfall throughout the year.

Rainfall in rainforests varies between 1 800 mm and 9 000 mm per year. What distinguishes a true rainforest is the distribution of rainfall throughout the year, as there is no dry season. A minimum of 100 mm of water falls as rain every month. If a rainforest has dry periods, these are usually short and unpredictable.

In many climates, rainwater vapour is transported to fall as rain in distant locations, but in rainforests almost 50% of rainfall comes from local evaporation. The warm, moist air floating in a rainforest forms a microclimate that does not allow much water to escape. Much of the rain that falls on a rainforest stays in the tops of the tallest trees. Some of it runs down their leaves and trunks to the lower trees and plants, but a good percentage evaporates and condenses in the form of small droplets that float in the humid atmosphere. Gentle, steady winds lift these droplets to higher layers of the atmosphere where they cool and form clouds. When enough of these droplets cool, they condense and fall as rain, starting the cycle all over again.

Despite the incredible lushness and great variety of flora, one of the peculiarities of rainforests is that the soils on which they stand are often poor in nutrients that can be absorbed by plant roots. Mineral nutrients have been washed away by heavy rains and high temperatures over thousands of years. To compensate for this deficiency, most tropical trees absorb as many nutrients as they can and store them in their inner layers. When a tropical tree dies, its nutrients decompose and go into the soil. Instead of being stored in the soil as they would be in a temperate forest, the nutrients are quickly absorbed by other living organisms.

The structure of a rainforest is different from other forest types because of the number of vegetation layers, known as strata. The lowest stratum is the undergrowth, composed of palms, herbaceous plants (such as wild ginger), blossoms and saplings. Only 2% of sunlight reaches this stratum, so the species that inhabit it have developed special adaptations to survive in low light. Many have a reddish colouration on the underside of their leaves to capture some of the scarce light that reaches the understorey. This colouration allows the plants to absorb light from a different spectrum than the lush green plants of the canopy. Above the undergrowth, but below the canopy, there are one or more strata of woody plants, such as large shrubs and medium-height trees.

The roof is the canopy, in which the canopy of trees forms a continuous layer that captures most of the rainwater and sunlight that reaches the forest. The height of the canopy depends on the region and the type of forest, and can range from 20 to 50 m. The lush green canopy explodes with life, and botanists have invented ingenious methods to access this mysterious ecosystem. Researchers use hot-air balloons, cables, catwalks, towers, sophisticated tree-climbing devices and even robots to study the millions of plants and animals that make their home in the canopy. Those studying this ecosystem also use large cranes.

dropped from helicopters into the heart of the forest. Suspended from the long, moving arm of the crane is a large gondola that functions as a mobile laboratory. Leaping from tree to tree, biologists collect samples, conduct experiments and observe life at the edge of the canopy.

The highest stratum of the rainforest is formed by emergent trees, those specimens that stand out above the forest canopy. The emergents, which do not form a continuous canopy, are usually the giants of the forest, reaching heights of 35-70 m and even more, with trunk widths of more than 2 m in diameter. Less than 1% of forest trees belong to the canopy or emergents. However, they are often so large that between them they account for most of the forest's stand, or biomass.

The beautiful tiered arrangement of the rainforest, including the continuous canopy layer, is regularly disturbed by natural events, such as falling trees.

The trees in a rainforest are often linked by vines, and when one tree falls it can knock down others, producing a domino effect. The clearing thus opened up in the forest canopy allows light to reach the forest floor. New plants and animals move into the area and begin to grow.

Other natural disturbances can open up even larger gaps. For example, along the hurricane-ravaged zone in the Caribbean Sea and the typhoon zone in the western Pacific, some forests are substantially altered when storms and high winds knock down hundreds of trees every few years. Scientists have found that these natural disturbances and the resulting forest regeneration are part of a natural process that produces healthy and diverse rainforests.

Rainforest ecosystems contain more plant and animal species than any other habitat in the world. Although their extent has expanded and contracted with climatic changes over the past millions of years, rainforests are generally among the oldest ecosystems on Earth. As a result of this continuity, rainforests are home to millions of different species, many of which are endemic, or unique to the rainforest habitat.

Although they contain numerous species, rainforests are remarkably uniform in their general appearance. Many trees have tall, slender trunks, which do not branch to near the crown. Kapok are supported by thick supports that can extend more than 10m. These supports provide the necessary support for rainforest trees, which are heavier in the canopy, as the nutrient-poor soils of the rainforest result in fragile, hollow roots. The bark of rainforest trees is usually thin and smooth. Notable exceptions are palms, which are common in some rainforests, yet absent from almost all other forest types.

Rainforest plants have many unique physical characteristics to take advantage of the particular habitat, or ecological niche, occupied by that species. Plants in the lower and middle levels of the forest, such as those related to the banana tree, tend to have especially large leaves to capture as much light as possible, i.e. the little light that has not already been intercepted by the canopy. These large leaves do not dry out as they would if they were high in the forest, where intense solar radiation creates a drier environment. These tendencies, however, can change when the environment is altered. Canopy trees change their shape over the course of their lives, depending on the surrounding environment. Leaves usually become smaller as the tree grows. In some cases, the leaves of juvenile specimens can be almost ten times larger than those of adults of the same species.

On the nutrient-poor forest floor, many plants such as *Astrocaryum sciophilum, which* belongs to the palm family and is about the size of a person, collect debris falling from other plants through their cup-shaped leaves to form their own reservoir of organic compounds.

Rainforests are also home to insectivorous plants, which collect a large part of their nutrients by capturing animals - especially insects - with their leaves. Among the insectivorous plants are the pitcher plants. The insects land on their tube-shaped leaves and then slide into a cavity, the pitcher, located in the stem and filled with digestive juices. There, the insect quickly dissolves and the nutrients it contains are stored by the plant.

Rainforests are often full of climbing plants, such as rattan palms. These thick, woody creepers (up to 25 cm in diameter) are often found attaching trees, climbing up to the canopy and hanging from the canopy. By climbing trees, these lianas expose their leaves and flowers to sunlight, birds and insects without expending the energy needed to build their own support systems. Epiphytes, such as mosses, bromeliads and orchids, grow on tree trunks or at the intersections of branches. Lacking permanent roots in the soil, epiphytes have to obtain their nutrients from other living plants, or by capturing water and organic matter as they fall to the forest floor. Bromeliads can store up to 38 litres of water in the tanks they form with their overlapping leaves. Many of them live harmoniously with their hosts, although some are less benign. Strangler plants, which start as epiphytes, germinate on trees in the forest canopy and send roots down to the ground. As they grow, these parasites envelop their hosts until they literally strangle them, at which point they become trees themselves.

FLYING OVER THE JUNGLE

Looks incredible, doesn't it, boy? Abel sat silently beside the pilot, still not quite recovered from the excitement of the early morning take-off from the packed earth of Simla, a little sleepy after a sleepless night, but inwardly anxious and curious, and gazing with wide-open eyes at the wonderful, ever-changing spectacle below them. In the distance they could see that strangely veiled sun, rising out of the mist like a red ball, advancing towards them, brighter and brighter, and then being absorbed again by the mist.

His astonished eyes beheld the peaks of the mountains of Trinidad, sometimes glowing reddish in the sunlight, sometimes seeming to float, shadowy and cold through the clouds; the ravines, from which the mist rose along the steep rock faces; jungles covered by that sea of mist more sensed than seen.

New peaks, new ravines, new clouds, new forests... mist, sun, mist, sun.

Suddenly, it seemed to Abel as if it were all a dream: to be sitting here next to the pilot, so high above the earth and even above the clouds. The monotonous noise of the engine, the vibration of the plane, echoing in his own body, incessantly, as if it would never end. At last the anguish he had experienced at the beginning left him. Nothing bad could happen to him, being at the side of his great friend.

Cautiously he put his hand on his companion's shoulder, who looked up at him smiling.

- That's wonderful, Jack!
- More beautiful than an Emilio Salgary novel or a football match, isn't it?
- Much more beautiful, much more beautiful!

This fantastic flight was like a birthday present from his friend. Mainly in the morning. He had always wanted to go to the jungle. Surely Julio Miguel Llanes López had come to an agreement with Jorge Luis Aneiros, Rector of the Spiritan University.

The aircraft, a very old Cessna 336-Skaymaster, but still in flying condition, served the William Beebe Tropical Research Station. This flight was to support an expedition to the rainforest.

Abel leaned back comfortably in his seat. The hum of the engine and its vibration made him sleepy. So did constantly looking outside. Abel fell into a light sleep, from which he was constantly startled awake every time his head leaned forward limply. Then, for a few seconds, he would see Jack's smiling face, the sun, the clear patches of mist, the tops of the mountains and the forests. To be awake and to dream were one and the same thing.

From time to time, his colleague's voice would come to rouse him from his slumber.

- The plains.

From the books he had read and his personal experiences of fieldwork in Cuba, he already knew those vast savannahs, stretching along the foothills of the mountain range, between the mountain rivers, interrupted by clumps of trees and bushes. But it was something very different to see it all with one's own eyes and in a different setting. Once a thin column of smoke rose, and he could see some people, small as beetles, walking through the undergrowth. Abel took the binoculars he had set up on the seat between him and his father to get a better look at them. Sometimes, like a dull mirror, a pond surrounded by reeds and bushes or a thin branch of a river glittered. Beyond that, for long stretches, one could only see flat, deep green expanses and small chains of hills. Ahead, again, people and animals walked one after the other in an irregular line. How many days would it take to cover a distance that the plane covered in a couple of hours?

After an hour's flight, they landed at Port Roy airfield, refuelled the tanks and took off again. The pilot had made his scientific and literary intentions for the flight clear.

When they had been flying again for ten minutes, the pilot handed Abel the map and, pointing to a spot, motioned with his head for him to look to the side.

- The Caroni River!

He read this name on the map and then followed the pilot's gesture with his eyes. He could not see the river itself, but he guessed that it had to be in the middle of the strip of jungle that stretched out below them.

He looked at his compass. They were flying west. Several times they crossed lesser streams, which for a few seconds shimmered through the jungle. No doubt tributaries of the Caroni.

The pilot flew the aircraft in a gentle curve, following the larger river course for a while, then descending steeply. Abel glanced at the altimeter: 170 metres above several small rivers. The plane was flying at an altitude of about 500 metres. Now it made the turns a little more sharply and descended a little further. But no matter how closely they watched, they did not see the expected smoke signal rising from the confusing array of jungles, clearings, swamps and rivers. The plane was rising again. The pilot went back to studying the map and the sketch.

- We have flown the correct course. Everything agrees with the sketch and the map of the indicated place. Let's do it again. Perhaps we have passed too quickly. Maybe they didn't have time to light the fire.

The pilot tried another tactic. He brought the plane up each time he made the turn, made the engine sound loudly, and then glided down, only to regain altitude when he came close to the range of hills. Once he raised his hands in a gesture of mute desperation and grumbled a few words between his teeth....

JUNGLE LANDING

They were flying southeast. A greyish wall was visible on the western side, and it seemed as if the view was becoming more and more veiled with each passing second. A few minutes later, almost without transition, they found themselves flying in the middle of a grey-green twilight.

At last, the bonfire! The reddish tongues of the flames, indicating to the aircraft crew the right direction to land in the makeshift glade, next to the intricate jungle.

The ship, which is pointing its propeller nose towards the makeshift glade next to the forest, a peat bog compacted by human effort, in the absence of modern technology suitable for such a rugged tropical landscape.

I am already a few metres from the jungle, a cathedral of branches and leaves projecting from the ground into the sky, the colossal hundred-year-old trees pricking the clouds with their crowns at intervals of thirty to forty-five metres the length of their trunks.

As he flew, he remembered what he had studied: "...Almost 90% of the animal species in the rainforest are insects, and most of them are beetles. A single tropical tree can harbour more than 150 species of beetles. Because they live high in the canopy, most of these beetles and other insect species have eluded scientists until recently, when technology has allowed access to the upper stratum. To this day, scientists are unsure how many animal species exist in the world, largely because they have identified only a small fraction of the millions (some say as many as 30 million) of insects that live in the rainforest.

Among the most fascinating of the insects already discovered are the leaf-cutter ants, remarkable because they cultivate their own food. These ants cut the leaves of certain plants and carry them to their underground nests, where they fertilise them with saliva. This careful process grows a particular fungus, which the ants collect and which is their only source of food.

In 15 km^2 of rainforest there can be up to 100 different species of mammals. These animals occupy every possible ecological niche, from burrows on the forest floor to the branches of emerging trees.

Most rainforest mammals are nocturnal (active at night) or crepuscular (active at dawn or dusk), and spend the hot part of the day sleeping. In fact, almost half of the mammals in the rainforests are bats - flying mammals notable for their nocturnal activity - some tropical mammals such as tapirs, agoutis and wild boars inhabit the ground, but most, like insects, prefer the high canopy of the forest canopy. The inhabitants of this canopy have evolved a whole panoply of fascinating features to survive in the branches of trees. Many tropical monkeys in Central and South America use their specialised branch-grabbing tails as a fifth limb when climbing, eating or even playing high in the canopy.

The three-toed sloth spends most of its life defenceless, hanging upside down in the branches of trees. To throw off predators, its movements are so slow that they are virtually undetectable, even to wary jaguars. The sloth has also developed a relationship with a tropical plant, which makes it even more elusive: although it has brownish fur, the sloth blends in with its natural green surroundings, because a particular type of green algae lives in its fur.

As with this type of algae and the sloth, many plants and animals in the rainforest depend on each other, even more so than in other ecosystems. For example, 90% of trees depend on animals to disperse their seeds. By comparison, in other forest types 50% or more of the trees use the wind to do so. These relationships between plants and animals are beneficial to both. Some animals protect plant species from predators while plants provide shelter, for example, many tropical plants such as snake trees have hollow structures in their stems that biting ants use as homes. In exchange for a place to live, the ants protect the plant, scaring off potential predators (climbing plants or hungry animals) as soon as they discover their presence.

In some cases, species are so dependent on each other that they cannot live separately. For example, all types of fig tree depend on one or more species of wasp, and vice versa. Without the task of pollination carried out by wasps, fig trees could not reproduce, and would face extinction; and without shelter for their eggs and larvae, a similar fate would await wasps.

Other animals that inhabit the Deltan and Trinidadian terrestrial space are: macaws, acures, chigüires, báquiros, tapirs, deer, jaguars, the ocelot, anteaters and anteaters, the cachicamo, iguanas, various types of snakes and many species of land, coastal and marine birds; among them macaws, parrots, moriches, jays, ducks, turkeys, paujíes, coitoras, gallinetas, herons, ciders, arucos and toucans.

As for the fish of the aquatic fauna we have the following: morocoto, cachama, zapoara, coporo, arapaima, guasa, lebranche, cazón and others; there are also babas, caimans, turtles, manatees and dolphins.

The rainforests also add a large number of exotic animals, such as parrots, toucans, hummingbirds, and monkeys, which are very valuable as pets.

An alternative in some forests is sustainable logging, where trees are carefully selected to minimise the impact on the ecosystem. Other communities harvest and sell biological products, such as valuable plant seeds (tagua nuts or Brazil nuts), while others engage in the production and marketing of medicines and drugs to strengthen and diversify their economies. Another increasingly popular alternative is the farming of beautiful tropical butterflies. In many rainforest communities, ecotourism (a type of tourism based on nature study and outdoor activities with minimal ecological impact) has proliferated as a means of attracting economic resources while preserving their fragile tropical ecosystem.

THE TRINIDAD JUNGLE SAFARI

I started to walk under a tunnel of vegetation, and as I approached a pond I plunged a stick into the mud. Muddy water gushed out of its mouth and a subterranean bellow shook the lake shore. The sound could hardly have come from an anaconda, the large South American boa constrictor we were looking for.

Not even a snake would own the big yellow tooth that I could barely make out deep in the darkness. "That settles it," said Jack Donne, leader of our animal-watching expedition for the University of Sancti Spíritus José Martí Pérez's Department of Forest Engineering to Trinidad Island. "It's a caiman."

The angry reptile eludes a snare. We were far back, in the region of the Caroni River, in the southwest of the island of Trinidad, in the rainforest near Venezuela. At that moment, about 20 copper-skinned Macushi Indians were watching, at a safe distance.

A black caiman or spectacled caiman *(Caiman crocodilus), a* close cousin of the caiman and crocodile, has two offensive weapons: first and obviously, its enormous jaws; and second, its immensely powerful tail.

Luckily, in this case we only had to deal with one end at a time. Having glimpsed that tooth, I knew which side to worry about. Working quickly, we cut saplings and sunk them deep into the mud to form a semi-circular stockade around the hole.

Without hesitation, Jack climbed inside and began splashing about, trying to see where the reptile lay. If the beast came out in a hurry, he would have to jump for the shore so as not to lose a leg. For my part, I was feeling quite nervous outside of what was at stake, wading in slowly as Professor Abel Hernandez manoeuvred a canoe.

Jack dangled a rawhide noose in front of the alligator's nose, waiting for the beast to pounce and stick its head in the noose. It roared and flailed so violently that the boat itself shuddered, but refused to be hoisted. With two forked sticks to keep the rope open, Jack slowly passed it over the alligator's snout. Enraged, the animal shook him off. Three times he pulled the rope taut and shook it off. Again, Jack pulled it into position. Suddenly he tightened the noose and the dangerous jaws were secured. Jack pushed a long sapling towards the den and, reaching inside, passed the rope over the branch, under the reptile's front legs, and hooked it tightly.

Then, pulling cautiously, he pulled out the rod and the black caiman or alligator, joining the hind legs and tail as it emerged violently from the silt.

The animal lay at our feet, muddy water sloshing around its jaws, its unblinking yellow eyes glaring malevolently at us.

It was only 10 feet long. Charlie and I were jubilant. Jack was less demonstrative. "Not bad," he said, "for a start, but there are bigger guys around here.

A while later, walking through the forest, we discovered a sloth hanging from some branches. The indolent sloth always sees the world upside down.

Scientists call this sloth *Bradypus,* a Greek word meaning slow-footed.

Suspended from its hooked claws, the sloth swung at a snail's pace through the jungle branches, as it does throughout the tropical forests of Central and South America.

In a whole day it will not be able to advance more than 50 metres. The creature takes life so calmly that its name is a symbol of laziness. This three-toed female hangs from a cecropia or guarumo (yagruma) tree by the Caroni River. It is one of the most conspicuous trees in the Neotropics.

The jungle of Trinidad offers a remarkable wealth of wildlife. The island provided us with ideal vantage points for our observations, films and incomparable photos of the tropical wildlife of this island.

Here there was not only thick rainforest, but also coastal streams and swamps, mangrove swamps, and inland savannahs. Each area had distinctive animal life. Jack Donne, the curator of reptiles at the Port of Spain Zoo, Charles Christenson, an experienced 16mm filmmaker, and I, the teacher and expeditionary team leader, had worked together once before in the Orinoco rainforest of Venezuela.

This time we were accompanied by Tim McPhee, one of the supervisors of the Port of Spain Zoo and a skilled animal keeper. Two days after we captured our yacaré. Jack, Charles and I were walking near a swampy creek with Teddy London, our host guide at the Saint Patrick outpost settlement.

Suddenly Teddy stopped and pointed to a footprint in the mud. "Anteater," he said laconically. We knelt down and examined the trail. The footprint of the great anteater is unmistakable. Its huge claws, used to tear up hills of cement-hard termites in search of soft grubs, fold back against hare's knuckles and leave tracks like fists dug into soft soil.

Catching an anteater by the tail is no easy task. Normally, the anteater is a harmless creature, but if it grabs an opponent with its powerful front paws, the victim has little chance of escaping its merciless claws.

As we sniffed through the undergrowth, there was a rustling sound and, a few metres away from us, a large, hairy form shuffled to its feet.

He galloped off, with Jack and Teddy at his heels. I ran out of the thicket and shouted to Charles to hurry up and get his camera. As we skirted the thicket, the anteater came lumbering in our direction.

I chased after it and, setting off again, slowly approached the animal. I wondered what to do? I caught it when Jack, about 100 metres away, shouted: "Grab its tail! Without thinking, I grabbed the animal by its tail end.

The creature turned around, reared up and charged at me with one of its front paws. I fell backwards and the aardvark once again lumbered away across the savannah, waving its tail like a banner. However, the pause was enough for one of Teddy's mounts to reach the beast with his lasso from the saddle. He swung his sisal lasso and threw it around the tamandua. In a matter of seconds, he had captured the animal, which growled in irritation.

Once the bear was securely tethered, we examined it closely. It was covered in dirty grey fur, and measured two metres from the tip of its furry tail to the tiny trunk at the end of a long, curved, toothless, proboscis-like snout. From here, its long, strap-like tongue, coated with sticky mucus, could extend even further to pick up ants and termites from anthills like a sticker. Its forelegs end in ten-centimetre claws.

On the way back to base camp, we watched as a tamandua came across a termite mound into which it cautiously inserted its long snout, licking the insects. A prehensile tail supports it, which hangs from it as it climbs a tree. The animal sometimes hangs by its tail.

On another occasion we spotted a matamata tortoise on the trail, whose rough shell and skin camouflage themselves to blend in with the soil of the ponds in the forest.

The next objective was in the Trinidadian rainforest. This time we headed west, crossing a steep hilly area, towards the delta of the Caroni River. This huge region is almost isolated from the rest of the country. Its scattered Acawi and Arecuna Indians still live relatively untouched by civilisation.

Only three men met us at the edge of the bumpy airstrip. The next morning we headed for a small tributary of the Caroni in a 40-foot canoe powered by an outboard motor.

"Nobody lives there," Bill told us, "so there must be a lot of wildlife. Among the four Acawi sent by Bill was the village chief. The next day, the chief led us through the jungle without hesitation, notching with his machete or bending bushes to prevent us from losing our way back.

Huge jungle trees towered 150 feet above our heads. From them hung orchids, bromeliads, epiphytic ferns, lianas and creepers, plants with large aerial roots.

Occasionally, on the forest floor, fallen yellow flowers were seen lying in a thick, colourful carpet. No large animals such as tapirs or collared peccaries appeared; but countless tiny creatures filled the sweltering twilight of the tropical rainforest with chirps, whistles and buzzes.

Three days later, we noticed our sloth reaching up to lick something on his hip. He was stroking a tiny baby. Still wet, the sloth cub must have been born only a few minutes earlier. The baby's colour was so similar to its mother's that, when it was dry, we could barely make it out among its mother's dense fur. From time to time it would slowly and laboriously climb up the length of its body to suck the milk from its mother's nipples. We watched the pair for two days. With her baby looking over her shoulder at us, the mother began to climb a tree. When we returned to the site an hour later, she was nowhere to be seen. Within a fortnight we had a large collection of animals for the Port of Spain National Museum and Gallery, from small insects, spiders, shells and birds to boas and capybaras, a giant relative of the guinea pig.

We often caught fireflies and bats at night. In the dark, the jungle was an eerie and mysterious place, full of sounds. We became accustomed to its incessant chorus, but the sound of a falling tree or an unfamiliar shriek would set my heart pounding.

Paradoxically, in the darkness we were able to find animals we would never have seen in the daytime. One night, as the shadows of night covered the jungle, pairs of eyes glowed at us: caimans lying submerged, their eyes just above the water.

Owls, owls, siju and bats are very common at night. As we walked through the black forest, a monkey turned to look at us. The reflections in the eyes of the beasts disappeared momentarily as they blinked, jaguar, ocelot, then vanished completely, and we heard a rustle of branches as they fled.

Walking as quietly as we could, we approached a thicket of bamboo, the stalks of which were swinging in the darkness. Jack shone the beam of his torch into the tangle of vegetation. "A good place for snakes," he said enthusiastically. "Make a noise from the other side and see if you can scare any animals away from me."

I made my way cautiously through the undergrowth, through the darkness and began to tap the bamboo canes with the back of a machete. Then the light from my torch fell on a small hole in the ground.

Cautiously I knelt down and looked in. From the depths shone three little lights. "Jack," I called softly, "here's something with three eyes!". It was beside me in three seconds. Beneath our two torches, a black, hairy spider the size of my open hand crouched at the bottom of the hole.

The eyes I had seen were only three of the eight that glowed on the top of its head. Menacingly it raised two forelegs topped with iridescent blue pads. Beneath, curved venomous fangs gleamed like brown hooks. "A beauty," muttered Jack. "Don't let it jump." He reached into his pocket for a vial of sample. I picked up a twig and pushed it gently through the hole. The hairy spider lashed out with its front legs. "Careful," Jack said. "If you get any of that hair off its body, it won't survive."

He handed me the vial. "I'll see if I can coax it out" He pressed his knife to the bottom of the gallery. The cavity crumbled and the tarantula suddenly ran backwards and landed directly into the collection vial. I quickly tightened the lid. Jack smiled and put the vial in his pocket. I later identified our catch as a specimen of *Avicularia avicularia, which is* very common in the jungle. Its venom is not fatal, but it can make a man who has been stung very sick.

Before leaving St Patrick's, Charles and I wanted to visit an Acawai Indian village. Once there, in a hut, we saw a baby parrot being fed with food in its beak. We stopped at small settlements every few kilometres to ask for animal specimens. Each village had its group of tame, pet parrots, which hopped over the eaves of the huts or walked irascibly through the clearing. Indians, like us, value parrots for the bright colour of their plumage and their ability to imitate human speech. Often, when we arrived, the birds would insult us in Acawai dialect. Adult parrots are difficult to catch and tame, so Indians collect budgerigars, parrots, Amazons and macaws from nests in the forest and raise them by hand in their homes.

In one village, a woman gave me a chick with big brown eyes, an absurdly large beak and some wispy feathers peeping through its bare skin, the so-called canyons. I couldn't refuse, but if I wanted to keep the attractive chick I had to learn how to feed it. The woman, laughing, showed me what to do. First I chewed some cassava bread. When the little bird saw this, it got tremendously excited, flapped its featherless wings and bobbed its head up and down in excitement.

I brought my face close to it and, without hesitation, the baby stuck its open beak between my lips. I pushed the chewed cassava down her throat with my tongue. This seemed very unhygienic to both the parrot and me, but the woman made it clear that there was no other method of raising a chick. Fortunately ours was quite advanced. A week later he was able to eat tender plantain on his own, freeing me from having to chew cassava every three hours.

I was impressed by the huge gaping mouth of an arapaima, a fish that puts up quite a fight when you try to catch it. A relative of salmon and herring, the arapaima, or pirarucu, is the largest known freshwater bony fish in all of South America.

In the Amazon basin, giants 15 feet long and weighing 400 pounds have been reported. Indians search for arapaima with nylon fishing poles and baited hooks; also with handmade harpoons or in collective attacks with bows and arrows.

The strength of the fish is prodigious. Anglers' tales of men being dragged from the river bank are frequent among the natives of the island, who know and appreciate it.

A very charismatic scavenger bird of the Trinidadian rainforest is the jungle condor *(Sarcoramphus papa),* also called royal condor, chom, royal crow, royal crow, king vulture, king zope, royal vulture or king vulture; a scavenger bird that cleans the tropical rainforests of carcasses.

Its head is a chromatic explosion, where the colours orange, yellow, pink, black, grey and white make their feast, contrasting with the black pupil, the white iris and the intense red eyeball. Its neck shows a grey tuft of down, while its body is greyish white.

Another very curious species is the playful capybara, the world's largest rodent, which organises aquatic fights in the waterways and impresses with its numerous herds.

Our last notable animal of Trinidad's fauna is found in the waters of the Caroni delta: the Antillean manatee, or sea cow, spends its life in rivers and brackish marshes, harmlessly eating its favourite herb.

It breathes air like a dolphin, its snout snorts and blows on the water. Being a mammal, its young suckle their mother's milk, sometimes emerging from the water with their calf cradled between their flippers. We searched hard for it until one day, we found our manatee. A large, fat animal, its blunt head adorned with a shaggy moustache. Tiny, sunken eyes stared at us, along with prominent nostrils.

THE MONKEYS OF TRINIDAD

Of the Trinidad monkeys with prehensile tails, the strangest in appearance, due to their exaggerated thinness, are the spider monkeys, which know how to use this appendage as a fifth hand, often hanging themselves from branches with it. The ability of these monkeys to climb with the help of their tails no doubt gave rise to the fable of the bridges they used to build with their bodies to cross rivers. The black coati *(Ateles paniscus)* is one of the best known species of this genus, characterised, among other things, by the lack of thumbs on its hands.

Spider monkeys or ateles, which owe their name to the excessive length of their limbs, have no thumb on their forelimbs, divergent noses and long, tactile, prehensile tails. They do not know how to escape from man, who pursues them for their skin as well as for their flesh. They do not form numerous bands and several species are known, but the one from the Trinidad jungle is *Ateles brisonii.*

Spider monkeys adapt easily to domestication and are very affectionate with their owners; I saw them on the shoulders of Indians in the villages we visited.

One day, loud cries ring out in the Trinidadian jungle: the monkeys howl. Darkness reigns inside the forest and there is a whiff of corrupted matter and actinomycetes. A few mosquitoes fly, an invisible toucan screams and large blue butterflies flutter among the branches.

We trudged panting through thorny bushes and lianas, tore our clothes into invisible hooks, impatiently circled the thorny caraguata, and scratched the skin of our faces and hands.

We are very close to the immediate herd of howler monkeys; soon their cries are heard above our heads, and the chaos of murmurs and howls resolves into the low tones of the old males, the high-pitched tones of the females and the shy tones of the young. Suddenly, all is quiet.

This very common species on the island is the howler monkey, another monkey with a handle tail that is distinguished by the great development of its hyoid, which forms a resonant box and gives its voice an incredible volume and sonority.

The tremendous bellowing of these monkeys, shouting in chorus, can be heard from a long distance at dawn and at sunset, and also at any time if rain threatens. Sometimes a single one bellows and then all the others respond.

The guariba *(Alouata seniculus), which* is an Amazonian species, distributed as far as the jungles of the Orinoco basin and the jungle of Trinidad, is a monkey in which both sexes have beautiful red-brown hair, with a golden yellow back.

This prodigious primate impressed me from the day of my arrival in the jungles of the island of Trinidad, and all the following days, I had heard the shrill cries of many monkeys resounding in the jungle at sunrise and sunset, but I never managed to see them. One morning, after breakfast, armed with my technical equipment, I entered the jungle, when that wild cry resounded in the thicket, which inflamed me with bellicose ardor. I hurried through the bushes and marched towards the place where the shriek resounded. Through the thicket, I hurried on towards the place where the noise sounded, and arrived where I intended without being observed.

Suspended in the heights of a tree in the forest canopy, these monkeys formed such a frightful concert that one might have suspected that all the wild beasts of the forest were engaged in mortal combat. Nevertheless, one could not but recognise that there was a certain order in the noise.

Suddenly the whole gang scattered around the tree would fall silent, and then one of the members, apparently the alpha male, would raise his intemperate voice and the howling would begin again.

The bony drum of the hyoid bone, which gives great sonority to its voice, widened from side to side. Sometimes its tones imitated the sound of other jungle animals.

The scene was very impressive, and even the most phlegmatic being would have been impressed by the gravity with which the bearded screamers sat facing each other.

OTHER TRINIDAD MAMMALS

In some coastal areas, such as south of the Caroni River basin, tropical mangroves abound. In the southern portion of the island of Trinidad, **jungle formations** grow up to 40 m high. A common tree there is the **araguaney** or yellow Ipe tree. There are felines like the jaguar, birds like the turpial, reptiles like the caiman and the anaconda, and rodents like the capybara, the largest capybara in the world!

The jaguar *(Felis onca concolor),* from all points of view, behaves like a fearsome predator. I was able to see that it moves with lightness and speed and jumps over six metres in length. In addition, it climbs deftly up trees and leaps to the ground, being unsurpassed by any other quadruped in swimming and fishing. It is known to attack fearsome alligators, tearing them to pieces and devouring them. Preferably it takes refuge on the edges of the jungle, especially along riverbanks and where there is an abundance of water.

The ocelot *(Leopardus pardalis) is* a species of carnivorous mammal of the Felidae family, much smaller than the jaguar, which is frequently found in the jungle of the island. It feeds on small mammals and birds, which it captures on the prowl.

The ocelot is a mammal belonging to the Felidae family that lives from the state of Texas (United States) to Argentina. The ocelot resembles a domestic cat and also a genet in appearance; it

measures between 55 and 100 cm in length and its tail ranges between 30 and 45 cm long. The body is robust and the legs are short. The colouring of the coat is very varied, to the point that no two ocelots are exactly alike. The background can be yellowish, grey, brownish or reddish and has black spots that are elongated on the front of the body and rounded on the back. The chest is somewhat lighter. The ocelot swims well and climbs trees very easily and frequently, although it also often descends to the ground. It hunts at night and feeds on birds, fish, snakes, lizards and small mammals. Little is known about reproduction; the time of mating is likely to vary according to latitude and the litter may consist of between one and four young. The skin of this animal is highly valued and has been hunted almost to extinction, and habitat loss due to conversion of forest to agricultural land has also contributed to the decline in the population of this species.
The nine-banded armadillo *(Dasypus novemcinctus)* is a species of placental mammal of the family Dasypodidae.

Armadillo, common name for a mammal related to anteaters and sloths. The nine-banded armadillo, also called the great mulita, tatú or toche, is distributed from northern Argentina to the southern United States. Other species native exclusively to South America are the six-banded armadillo or gualacate, the three-banded armadillo or quirquincho bola and the giant armadillo or tatú carreta. The size of these animals varies from 15 cm, which is the length of the smallest species, to 1 m in length, excluding the tail, in the case of the giant armadillo.

The armadillo is an armoured mammal; the armour of this animal consists of a mosaic of small bony plates that develop in the lower layer of the skin, or dermis, and are covered with horny epidermis. It constitutes a protection system against predators; in some species even the tail is protected. In all species except the giant armadillo, the plates form a one-piece shield over the shoulders and another over the hindquarters. The middle part of the animal's body is covered with plates that form transverse bands that are hinged together, i.e. they are connected by a soft skin. This allows them to roll up into a fully armoured ball and cover the only unprotected area of the body: the abdomen.

They are robust-bodied animals, with short, muscular legs that allow them to move fairly quickly. The limbs have claws similar to nails and the armadillo uses them to obtain food or to dig burrows. They are nocturnal and feed on insects, worms, small vertebrates and sometimes carrion. It can live in jungles as well as in open areas. The female usually has a large litter of up to 12 offspring in some cases; however, a small number of species give birth to only one offspring. Armadillo meat is edible. In South America, fossil remains of extinct groups close to armadillos, the glyptodontids *(see Glyptodon)*, have been found in Pleistocene geological strata. Hunting and habitat destruction have endangered the survival of several species, such as the giant armadillo and the three-banded armadillo.

The common vampire *(Desmodus rotundus)* is a species of microchiropteran bat of the phyllostomatid family, a common name for bats that feed on the blood of other animals: the true or common vampire, the white-winged vampire and the hairy-legged vampire. Their range extends from Mexico to South America, covering Chile, Argentina and the Margarita and Trinidad islands off the north coast of Venezuela.

Vampires weigh between 15 and 50 g and are only 6.5 to 9.3 cm long. They are grey, brown or reddish-brown in colour. The edges and wing tips of the white-winged vampire, as the name suggests, are white. Vampires have fewer teeth than other bats; the number of teeth is between 20 and 26, depending on the species. They have large, sharp incisors for piercing prey flesh and an anticoagulant in their saliva that keeps the blood flowing while they feed.

Vampires live in both arid and humid regions and in both forests and open terrain. They take refuge in caves, mines, abandoned wells, hollow logs and abandoned buildings, usually alone or in small groups of 12 or fewer individuals; however, the common vampire sometimes forms colonies of up to 2 000 animals. Common vampire colonies of 50 or more individuals are usually divided into smaller social units consisting of 8-20 adult females, one adult male and the young. Females have a single brood once or twice a year. Vampires have been known to live up to 27 years in captivity.
The common vampire feeds on the blood of other mammals, such as horses, pigs, donkeys and cows. They also feed on domestic fowl, such as turkeys and chickens. The hairy-legged, white-winged vampires live primarily on birds. All have excellent eyesight and

sensitive hearing and are very agile on the ground, where they use their powerful hind legs to crawl, run and leap into the air. To feed, they begin by licking the victim's skin to soften it. They then cut the hair or feathers before piercing the skin with their sharp incisors and sucking out the blood. They can drink about 20 ml per night, almost half their weight.
Sometimes the wounds caused by vampires become infected; in addition, these bats can transmit rabies. In their attempts to eliminate vampires that feed on livestock, ranchers kill large numbers of useful bats. One method of vampire control that does not endanger other bats or livestock is to apply vaseline containing anticoagulant to the backs of live vampires, so that other animals in the colony ingest the poison in the course of mutual social cleansing behaviour. Vaseline is also applied to the backs of domestic animals so that the bat absorbs it along with the blood.

The anticoagulant proteins in vampire saliva are considered to be about twenty times more potent than any known anticoagulant. It has been used to develop a drug called *Draculin, which is* used to treat heart attacks and strokes in humans.

EPILOGUE

The jungles of the island of Trinidad provided us with unique opportunities to observe its wildlife, its fronds showing animals of an insular territory of the Antilles, which treasures the most impressive, rich and valuable faunal heritage of the entire insular Caribbean.

The island is also an unparalleled scenery, with unique and paradisiacal landscapes, varied relief, offering very different scenarios for the various terrestrial ecosystems.

Its people are very kind, gentle and affable, making it a very special and productive area for ecotourism, wildlife watching, bird watching and conservation of animal biodiversity.

BIBLIOGRAPHY

Adam Woolfitt/Woodfin Camp and Associates, Inc. 2009. *Pluvisilvas.* Microsoft Student Encarta 1993-2008. Microsoft Corporation.

Bates, M. 1966. *The land and fauna of South America.* LIFE Nature Collection in Spanish. Published by Offset Multicolor, S. A. Mexico, D. F., 200 pp.

Bright, M. 2005. *Sudamérica Salvaje.* Ediciones B, Grupo Zeta. BBC. London, 256 pp.

Carlton Centre San Fernando. 2016. *Trinidad and Tobago NEWSDAY. No. 8319. Friday, July 1, 2016.* Published by Daily News Limited. Port Spain, 88 pages.

Hernández-Muñoz, A. 2015-2016. *Cuaderno de campo de mis campañas zoológicas en la Isla de Trinidad (*unpublished).

Marín Rodríguez, C. J. 1996. *Conservationist issues.* Editorial Litógrafos Asociados. Venezuela, Estado Monagas, Tucupita, 79 pp.

Mep. 2013. *Discover Trinidad & Tobago.* Media & Editorial Projects. Port Spain, 132 pp.

Rodríguez de la Fuente, F. 1981. *South America. Enciclopedia Salvat de la Fauna, Volume 8.* SALVAT S. A. de Ediciones. Barcelona, 713 pp.

TDC. 2016. *Trinidad & Tobago. Where the World Meets.* Trinidad and Tobago Convention Bureau. Tourism Development Company Limited. Port Spain, 3 pp.

Microsoft Student Encarta. 2009. *William Beebe Field Station.* Microsoft Student Encarta 1993-2008. Microsoft Corporation.

Microsoft Student Encarta. 2009. *Republic of Trinidad and Tobago.* Microsoft Student Encarta 1993-2008. Microsoft Corporation.

White, G. 2023. *IBAs* of *Trinidad and Tobago.* Trinidad and Tobago Field Naturalist Club. Port Spain, pp. 309-313.

Printed by Books on Demand GmbH, Norderstedt / Germany